12 Steps to Intelligent Process Improvement (IPI)

Restored Legacy Quick Reference Guide

Companion to Feng Shui for Intelligent Process Improvement: Legacy Principles of Structured Software Development

Bringing balance and harmony to the workplace through Intelligent Process Improvement.

By V. Quan Lee and J. Kay

12 STEPS
TO INTELLIGENT PROCESS IMPROVEMENT (IPI)
Restored Legacy Quick Reference Guide

Companion to Feng Shui for Intelligent Process Improvement: Legacy Principles of Structured Software Development

♦ ♦ ♦

By V. Quan Lee and J. Kay

Intelligent Systems Press

Configuration control, metrics, structured SDLC gates, process champions, and disciplined accountability are presented as practical tools for restoring balance and harmony in the workplace, not as museum artifacts.

Any references to methodologies, corporate behaviors, binders, shared drives, templates, status meetings, and consulting rituals are used for commentary, education, satire, and historical process-improvement discussion.

Copyright and Publishing Notice

Editing by: Brian Loveless
Web Content Design by: Erik Victor Jones
Content Editing and Layout by: Emma Victoria Lee

ISBN: 979-8-9958725-6-6

This restored legacy quick reference guide is prepared by **Intelligent Systems Press** as a companion to *Feng Shui for Intelligent Process Improvement: Legacy Principles of Structured Software Development*. It preserves the original IPI quick-reference spirit while refining the language, structure, usability, and visual direction for a publishing-ready guide.

This guide intentionally retains the vocabulary of legacy software process engineering, including requirements, signoffs, process libraries, audits, and related terminology.

TABLE OF CONTENTS

Author's Note to the Restored Legacy Edition

This quick reference guide is intentionally written as a restored legacy companion to Feng Shui for Intelligent Process Improvement: Legacy Principles of Structured Software Development. It preserves the old-school software process improvement mindset: requirements, signoffs, process libraries, audits, metrics, process champions, change control, and disciplined accountability.

The purpose of this restored guide is not to modernize IPI into the language of newer delivery trends. Its purpose is to keep the original IPI spirit alive: process improvement for the people and by the people, with enough structure to reduce chaos and enough common sense to avoid turning the workplace into a shrine to paperwork.

The Feng Shui comparison remains central. A software workplace, like a physical environment, must be arranged so energy can flow. In IPI, that energy moves through people, artifacts, decisions, handoffs, measurements, approvals, and lessons learned. When those elements are misplaced, frustration grows. When they are arranged with purpose, the workplace becomes more balanced, more accountable, and more humane.

Use this guide as a practical field reference. It is meant to be read quickly, marked up, carried into meetings, and used by the real workers who know where the process breaks before the dashboard admits it.

About the Authors

V. Quan Lee and J. Kay bring together two complementary views of software process improvement: the disciplined world of quality assurance, audit readiness, structured SDLC control, and formal process accountability; and the practical world of business analysis, project delivery, customer communication, training, and real-worker execution.

Their shared perspective is grounded in the belief that process improvement should not be reserved for elite methodology specialists, expensive consulting teams, or organizations large enough to finance endless training classes. Process improvement belongs to the people who must live with the process every day.

This restored quick reference guide honors that original point of view by preserving the humor, frustration, structure, and workplace realism that shaped Intelligent Process Improvement. It keeps the focus where IPI has always placed it: on the real workers of IT and the practical steps required to bring harmony, clarity, accountability, and balance into the software development workplace.

About the Publisher

Intelligent Systems Press publishes practical works at the intersection of software development, process improvement, technology leadership, and enterprise operating discipline.

The imprint exists to preserve and advance ideas that help organizations bring clarity to complex work. Its publishing philosophy is simple: useful books should give readers a framework, a language, a set of tools, and the confidence to improve the systems around them.

This restored legacy quick reference guide is part of a broader body of work on Intelligent Process Improvement and Process Intelligence. It is designed to stand beside Feng Shui for Intelligent Process Improvement: Legacy Principles of Structured Software Development as a compact companion for readers who want the IPI principles in a practical, portable format.

This restored legacy quick reference guide is part of the Intelligent Process Improvement and Process Intelligence publishing series. It replaces the original IPI LLP / Ying Yang Enterprises contact block with the updated publisher identity used in Feng Shui for Intelligent Process Improvement: Legacy Principles of Structured Software Development.

Published by Intelligent Systems Press

This restored legacy quick reference guide is part of the Intelligent Process Improvement and Process Intelligence publishing series. It uses the updated publisher identity from Feng Shui for Intelligent Process Improvement: Legacy Principles of Structured Software Development while preserving the nostalgic legacy IPI concept.

Intelligent Systems Press publishes practical works at the intersection of software development, process improvement, technology leadership, and enterprise operating discipline.

How to Use This Quick Reference

This is not a replacement for the full restored legacy manuscript. It is the pocket companion: a quick, practical field guide for workers, managers, process champions, auditors, trainers, and project leaders who need the IPI concept in a faster format.

The guide should be read in the old-school spirit of process implementation: print it, mark it up, carry it to a meeting, use it during a process workshop, or place it beside the project binder that everyone promises they will open before the next audit.

Legacy Reading Note

This guide does not modernize IPI into a trend. It keeps the nostalgic, structured, people-centered process-improvement concept intact: Feng Shui creates the metaphor, IPI creates the operating discipline, and the real workers make the process come alive.

IPI Mission Statement

The purpose of Intelligent Process Improvement is to intelligently select and implement incremental improvements that measurably enhance an organization's processes by creating harmony, clarity, accountability, and empowerment for the real workers in IT.

At the end of the day, IPI is about taking the frustration out of the software development process.

Figure 1 — Legacy IPI Operating Compass

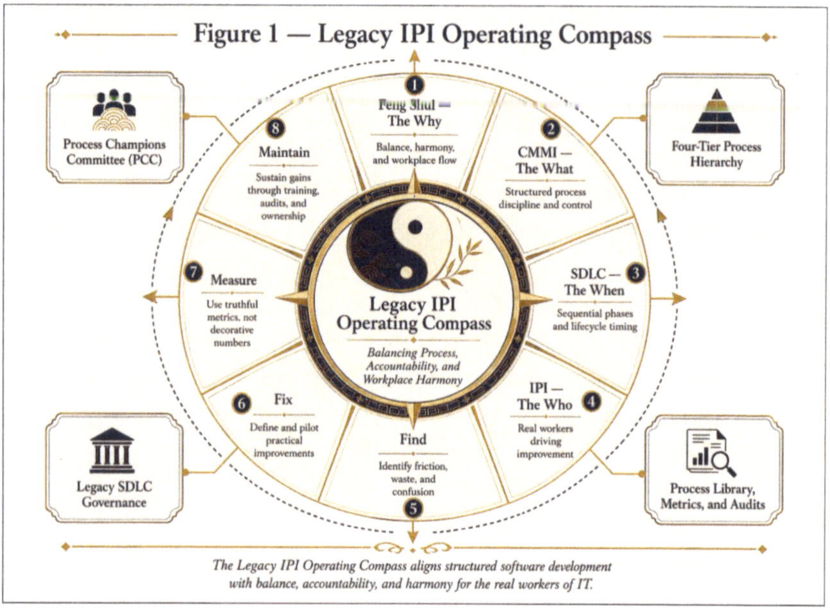

The Legacy IPI Operating Compass aligns structured software development with balance, accountability, and harmony for the real workers of IT.

The 12 Steps at a Glance

Step	Legacy Action	Quick Meaning
1	**Begin with the people and the principle**	Process improvement is for the people and by the people.
2	**Define your own hybrid process approach**	Look at best practices and create what works for your environment.
3	**Make process improvement a legitimate project**	Give IPI a budget, timeline, tools, resources, and a project manager.
4	**Inventory the as-is process honestly**	Admit the current reality without punishing the people carrying it.
5	**Make the process visible**	Draw the before picture, chart progress, and expose the flow of work.
6	**Train, display, and reinforce constantly**	Infuse process improvement into the culture through training and visual reminders.
7	**Empower the Process Champions Committee**	Give the PCC centralized authority to govern, arbitrate, audit, and sustain.
8	**Hold everyone accountable to the PCC**	Executives, managers, and workers all live under the process discipline.

9	**Improve in short controlled intervals**	Use Find, Fix, Measure, Maintain in practical increments.
10	**Keep processes flexible enough to serve business**	Do not let rigidity lock the organization out of opportunity.
11	**Measure what matters, including harmony**	Set quantified goals for quality, frustration, empowerment, balance, and results.
12	**Purge process overload**	Retire processes that no longer serve the work, the worker, or the workplace.

IPI 12-Step Quick Card

Step	Legacy Action	Ask Yourself
1	**Begin with the people and the principle**	Have the real workers been heard before the process is defined?
2	**Define your own hybrid process approach**	Does this process fit our actual environment, culture, size, and work style?
3	**Make process improvement a legitimate project**	Is IPI supported with ownership, schedule, resources, tools, and visible sponsorship?
4	**Inventory the as-is process honestly**	Are we documenting the real current process without blame, fear, or fictional cleanup?
5	**Make the process visible**	Can the workflow be drawn, understood, inspected, and improved by the people using it?
6	**Train, display, and reinforce constantly**	Has the process been taught, displayed, repeated, and refreshed enough to become part of the culture?
7	**Empower the Process Champions Committee**	Does the PCC have real authority to govern, arbitrate, audit, approve, and improve processes?
8	**Hold everyone accountable to the PCC**	Do executives, managers, and real workers all follow the same process discipline?
9	**Improve in short controlled intervals**	Are we using Find, Fix, Measure, Maintain instead of attempting one massive rollout?
10	**Keep processes flexible enough to serve business**	Can the process bend for legitimate business needs without collapsing into chaos?
11	**Measure what matters, including harmony**	Are we measuring clarity, quality, frustration, empowerment, accountability, and flow — not just cost and schedule?
12	**Purge process overload**	What outdated process, template, approval,

Step	Legacy Action	Ask Yourself
		checklist, or artifact should be retired before it blocks the work?

Quick Use Instructions

Use this card at the beginning of an IPI workshop, PCC meeting, process review, audit preparation session, or project kickoff. The purpose is not to answer every question perfectly. The purpose is to expose where the process is out of balance.

IPI Reality Check

If the team cannot answer the question beside a step, that step probably needs attention before the organization adds another template, schedules another meeting, or blames another real worker.

30-Minute IPI Workshop Agenda

5 minutes — Identify the process pain point

5 minutes — Name the real workers affected

5 minutes — Draw the current flow

5 minutes — Identify missing artifacts, owners, or approvals

5 minutes — Select one fix to pilot

5 minutes — Assign owner, metric, and follow-up date

Step 1

Begin with the People and the Principle

Process improvement is for the people and by the people. Processes should bring harmony and balance to the workplace.

♦ ♦ ♦

Legacy Purpose

Step 1 establishes the heart of IPI. The process must not begin as an executive command, a consultant package, or a borrowed methodology. It must begin with the belief that the people doing the real work have process intelligence because they live inside the work every day.

Feng Shui-to-Process Meaning

In Feng Shui, a room must be arranged around the life that occurs inside it. In IPI, a process must be arranged around the people who must perform it. When the people are ignored, the workplace energy is blocked before the first template is issued.

Practical IPI Actions

- Name the real workers affected by the process.
- Ask what frustrates them before defining what they must follow.
- Identify where work currently loses energy, time, quality, or trust.
- Treat worker feedback as process evidence, not complaining.
- Define the improvement purpose in plain language.

IPI Reality Check

If the people who perform the process cannot explain why it exists, the process is already drifting toward ceremony.

Pocket Checklist

Quick Check	Yes / No	Notes / Owner
Real workers have been identified.	☐ Yes ☐ No ☐ Partial	
The purpose is explained in plain language.	☐ Yes ☐ No ☐ Partial	
Current frustration has been captured.	☐ Yes ☐ No ☐ Partial	
The improvement is tied to workplace harmony.	☐ Yes ☐ No ☐ Partial	

Step 2

Define Your Own Hybrid Process Approach

Look at best practices, methods, and models that have been tried, then define a hybrid one that works best for your environment.

♦ ♦ ♦

Legacy Purpose

IPI does not ask an organization to pledge allegiance to a single method. It asks the organization to study available practices, understand its own culture, and assemble the smallest practical set of controls that will work.

Feng Shui-to-Process Meaning

Feng Shui does not place every object in every room the same way. Placement depends on purpose, space, flow, and energy. IPI uses the same idea: choose practices that fit the environment, not practices that merely impress outsiders.

Practical IPI Actions

- Review existing methods without worshiping them.
- Identify what has worked in your environment before.
- Borrow practices that solve real problems.
- Reject practices that create weight without value.
- Document why each selected practice belongs.

IPI Reality Check

A hybrid process is not a sloppy process. It is an intentional process built from useful parts.

Pocket Checklist

Quick Check	Yes / No	Notes / Owner
Best practices were reviewed.	☐ Yes ☐ No ☐ Partial	
The selected approach fits company size and culture.	☐ Yes ☐ No ☐ Partial	
Unnecessary practices were excluded.	☐ Yes ☐ No ☐ Partial	
The hybrid model is documented and explainable.	☐ Yes ☐ No ☐ Partial	

Step 3

Make Process Improvement a Legitimate Project

Make process improvement a legitimate project with a budget, timeline, tools, assigned resources, and a project manager.

♦ ♦ ♦

Legacy Purpose

Process improvement fails when it is treated as extra work performed by already exhausted people in the margins of their real assignments. A serious IPI effort needs governance, time, tools, ownership, and visible sponsorship.

Feng Shui-to-Process Meaning

A balanced room requires intentional arrangement. A balanced process initiative requires intentional project structure. Without placement, every improvement task becomes hallway work.

Practical IPI Actions

- Create an IPI project charter.
- Assign a project manager or process improvement lead.
- Define schedule, scope, milestones, and decision rights.
- Provide a process repository or portal.
- Give workers time to participate without punishing their delivery commitments.

IPI Reality Check

A process improvement effort with no budget, no schedule, and no owner is not an initiative; it is a wish with meeting minutes.

Pocket Checklist

Quick Check	Yes / No	Notes / Owner
IPI project charter exists.	☐ Yes ☐ No ☐ Partial	
Project owner is named.	☐ Yes ☐ No ☐ Partial	
Timeline and milestones are visible.	☐ Yes ☐ No ☐ Partial	
Tools and repository are identified.	☐ Yes ☐ No ☐ Partial	

Step 4

Inventory the As-Is Process Honestly

Take full inventory of your as-is processes and do not punish anyone for admitting weaknesses, shortcuts, mistakes, or workarounds.

♦ ♦ ♦

Legacy Purpose

The as-is process must be discovered honestly. Many organizations ask for current-state documentation and then punish the people who admit that the real process is informal. IPI rejects that trap. Truth comes first.

Feng Shui-to-Process Meaning

In Feng Shui, imbalance cannot be corrected until the environment is observed as it actually exists. IPI observes the current workflow without pretending that every missing artifact is a moral failure.

Practical IPI Actions

- Interview workers without blame.
- List documented and undocumented workflows.
- Identify shortcuts that exist because the official process is unusable.
- Capture missing owners, approvals, and artifacts.
- Record pain points as improvement candidates.

IPI Reality Check

If the team has to invent documents to prove it has a process, the organization has just documented fiction.

Pocket Checklist

Quick Check	Yes / No	Notes / Owner
As-is process is documented honestly.	☐ Yes ☐ No ☐ Partial	
Informal practices are identified.	☐ Yes ☐ No ☐ Partial	
Workers are not punished for revealing gaps.	☐ Yes ☐ No ☐ Partial	
Pain points are prioritized.	☐ Yes ☐ No ☐ Partial	

Step 5

Make the Process Visible

Draw a picture for everyone to see. Create a before picture of the process, make it visible, and use it to highlight progress.

♦ ♦ ♦

Legacy Purpose

Process mapping turns invisible confusion into shared evidence. Once the workflow can be seen, people can stop arguing from memory and begin improving from a common picture.

Feng Shui-to-Process Meaning

Feng Shui values placement and flow. A process map shows where the flow is blocked, where energy leaks, where roles collide, and where work enters rooms with no exit.

Practical IPI Actions

- Draw the current-state workflow.
- Show handoffs between roles and departments.
- Identify entry and exit points.
- Mark gaps, bottlenecks, loops, and unowned steps.
- Post or publish the map so everyone can inspect it.

IPI Reality Check

A process that cannot be drawn probably cannot be followed consistently.

Pocket Checklist

Quick Check	Yes / No	Notes / Owner
Current-state map exists.	☐ Yes ☐ No ☐ Partial	
Handoffs are visible.	☐ Yes ☐ No ☐ Partial	
Bottlenecks are marked.	☐ Yes ☐ No ☐ Partial	
Progress can be compared to the before picture.	☐ Yes ☐ No ☐ Partial	

Step 6

Train, Display, and Reinforce Constantly

Continuously indoctrinate and infuse process improvement into the culture through training and visual displays.

♦ ♦ ♦

Legacy Purpose

The word indoctrinate carries the original satire, but the principle is serious: process must be taught repeatedly. A process announced once is not a process learned.

Feng Shui-to-Process Meaning

Feng Shui creates harmony through repeated awareness of environment and placement. IPI creates process harmony by keeping expectations visible, teachable, and reinforced.

Practical IPI Actions

- Create role-based training.
- Use visual process displays and quick-reference cards.
- Review process updates in team meetings.
- Train new employees before they inherit bad habits.
- Refresh training when audits or metrics reveal confusion.

IPI Reality Check

If people only hear about the process during an audit, the process is not part of the culture.

Pocket Checklist

Quick Check	Yes / No	Notes / Owner
Training plan exists.	☐ Yes ☐ No ☐ Partial	
Visual aids are available.	☐ Yes ☐ No ☐ Partial	
New employees receive orientation.	☐ Yes ☐ No ☐ Partial	
Process updates trigger refresher training.	☐ Yes ☐ No ☐ Partial	

Step 7

Empower the Process Champions Committee

Give the PCC centralized power to govern, arbitrate, audit, and improve every process and project.

♦ ♦ ♦

Legacy Purpose

The PCC is the IPI operating body. It gives process improvement a home, a voice, a structure, and a sustaining mechanism. It prevents process from belonging only to management or only to consultants.

Feng Shui-to-Process Meaning

In Feng Shui terms, the PCC keeps the workplace energy aligned. It watches for blocked flow, unbalanced accountability, excessive burden, and process objects placed where no worker can reach them.

Practical IPI Actions

- Create a PCC charter.
- Include standing members and rotating real-worker representatives.
- Define authority over process assets, audits, waivers, and improvements.
- Review process suggestions and pain points.
- Report findings to executive leadership.

IPI Reality Check

A committee without authority is a suggestion box with chairs.

Pocket Checklist

Quick Check	Yes / No	Notes / Owner
PCC charter is approved.	☐ Yes ☐ No ☐ Partial	
Members are named.	☐ Yes ☐ No ☐ Partial	
Authority is defined.	☐ Yes ☐ No ☐ Partial	
Meeting rhythm is established.	☐ Yes ☐ No ☐ Partial	

Step 8

Hold Everyone Accountable to the PCC

Enforce that everyone, including executive management, is accountable to the Process Champions Committee.

♦ ♦ ♦

Legacy Purpose

IPI accountability is not punishment. It is balance. If only workers follow the process while executives bypass it, the process becomes a class system. IPI requires all levels to respect the same operating rules.

Feng Shui-to-Process Meaning

Harmony requires balanced energy. A workplace cannot be balanced when authority can ignore the process but labor must obey it. Executive waivers, documented decisions, and visible approvals restore fairness.

Practical IPI Actions

- Require executive process waivers for exceptions.
- Document decisions that bypass normal flow.
- Make accountability visible and role-based.
- Track repeated exceptions for process review.
- Ensure managers model the process they expect others to follow.

IPI Reality Check

Nothing destroys process credibility faster than leadership treating the process as optional.

Pocket Checklist

Quick Check	Yes / No	Notes / Owner
Executive accountability is documented.	☐ Yes ☐ No ☐ Partial	
Waivers are tracked.	☐ Yes ☐ No ☐ Partial	
Exceptions are reviewed.	☐ Yes ☐ No ☐ Partial	
Managers follow the same process discipline.	☐ Yes ☐ No ☐ Partial	

Step 9

Improve in Short Controlled Intervals

Make improvements in short iterations: 30 days for major improvements and 2 weeks for minor improvements using Find, Fix, Measure, Maintain.

♦ ♦ ♦

Legacy Purpose

IPI avoids the grand process rollout that takes a year to explain and collapses after the first project crisis. Improvements should be small enough to pilot, measure, and adjust.

Feng Shui-to-Process Meaning

In Feng Shui, harmony is restored through deliberate adjustments, not by throwing all the furniture into the hallway. IPI improves in manageable movements so the workplace can absorb the change.

Practical IPI Actions

- Find a specific friction point.
- Fix it with a practical pilot.
- Measure before and after evidence.
- Maintain the improvement through training, library updates, and audits.
- Repeat the cycle.

IPI Reality Check

A small process improvement that sticks is more valuable than a massive rollout that becomes a legend of failure.

Pocket Checklist

Quick Check	Yes / No	Notes / Owner
Improvement interval is defined.	☐ Yes ☐ No ☐ Partial	
Pilot scope is manageable.	☐ Yes ☐ No ☐ Partial	
Metrics are selected before rollout.	☐ Yes ☐ No ☐ Partial	
Maintenance steps are planned.	☐ Yes ☐ No ☐ Partial	

Step 10

Keep Processes Flexible Enough to Serve Business

Keep processes and methods flexible so you are not so rigid that you lock yourself out of business opportunities.

♦ ♦ ♦

Legacy Purpose

Legacy process discipline does not require rigidity. A good process protects quality while allowing sensible tailoring. IPI values discipline that can bend without breaking.

Feng Shui-to-Process Meaning

Feng Shui balances fixed structure with living flow. A process should provide enough boundaries to create stability and enough flexibility to respect project reality.

Practical IPI Actions

- Define required, optional, and project-defined steps.
- Use tailoring rules instead of informal bypasses.
- Create time-limited waivers for legitimate exceptions.
- Review whether recurring exceptions mean the process needs revision.
- Protect business opportunity without abandoning accountability.

IPI Reality Check

Flexibility without control becomes chaos. Control without flexibility becomes bureaucracy.

Pocket Checklist

Quick Check	Yes / No	Notes / Owner
Tailoring rules exist.	☐ Yes ☐ No ☐ Partial	
Required steps are clear.	☐ Yes ☐ No ☐ Partial	
Waivers are time-limited.	☐ Yes ☐ No ☐ Partial	
Recurring exceptions trigger review.	☐ Yes ☐ No ☐ Partial	

Step 11

Measure What Matters, Including Harmony

Set quantified measurements for every goal you wish to achieve. Measure harmony and balance, not just dollars and deadlines.

♦ ♦ ♦

Legacy Purpose

IPI expands measurement beyond schedule and cost. It asks whether process improvement reduces frustration, improves empowerment, supports quality, and restores workplace balance.

Feng Shui-to-Process Meaning

Feng Shui is concerned with the quality of the environment. IPI asks the workplace to measure whether the process environment is healthier after the improvement than before it.

Practical IPI Actions

- Define process performance measures.
- Measure rework, defects, delays, and cycle time.
- Survey frustration, clarity, empowerment, and cross-team trust.
- Make metrics visible to workers and leadership.
- Use data to adjust the process, not to decorate status reports.

IPI Reality Check

A metric that cannot trigger a decision is office decoration with decimal points.

Pocket Checklist

Quick Check	Yes / No	Notes / Owner
Each goal has a metric.	☐ Yes ☐ No ☐ Partial	
Harmony or frustration is measured.	☐ Yes ☐ No ☐ Partial	
Metrics are visible.	☐ Yes ☐ No ☐ Partial	
Results lead to decisions.	☐ Yes ☐ No ☐ Partial	

Step 12

Purge Process Overload

Purge processes and avoid process overload. Never hold onto processes out of tradition.

♦ ♦ ♦

Legacy Purpose

A legacy process environment must also know when to retire what no longer serves. Old templates, outdated approvals, retired artifacts, and excessive checklists can clog the workplace like furniture blocking a doorway.

Feng Shui-to-Process Meaning

Feng Shui teaches that clutter blocks energy. IPI teaches that stale process assets block work. Purging process overload is not anti-discipline. It is disciplined housekeeping.

Practical IPI Actions

- Review every process on a defined cycle.
- Retire or revise processes that no longer add value.
- Archive outdated templates clearly.
- Ask real workers which forms create burden without benefit.
- Track whether purging improves clarity and flow.

IPI Reality Check

A process kept only because it has always existed has become institutional clutter.

Pocket Checklist

Quick Check	Yes / No	Notes / Owner
Review dates are assigned.	☐ Yes ☐ No ☐ Partial	
Outdated processes are retired.	☐ Yes ☐ No ☐ Partial	
Process library is cleaned.	☐ Yes ☐ No ☐ Partial	
Workers confirm usability improved.	☐ Yes ☐ No ☐ Partial	

Figure 2 — The Legacy Workplace Balance Model

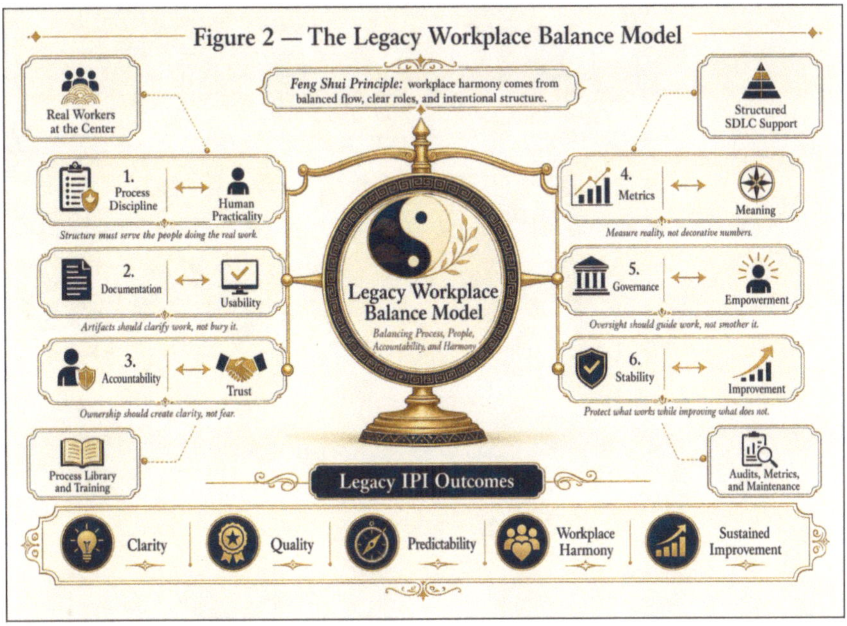

Use this model during Step 1, Step 10, and Step 11 to remind the team that process discipline, human practicality, metrics, governance, stability, and improvement must remain balanced.

Figure 3 — 90-Day Legacy IPI Implementation Roadmap

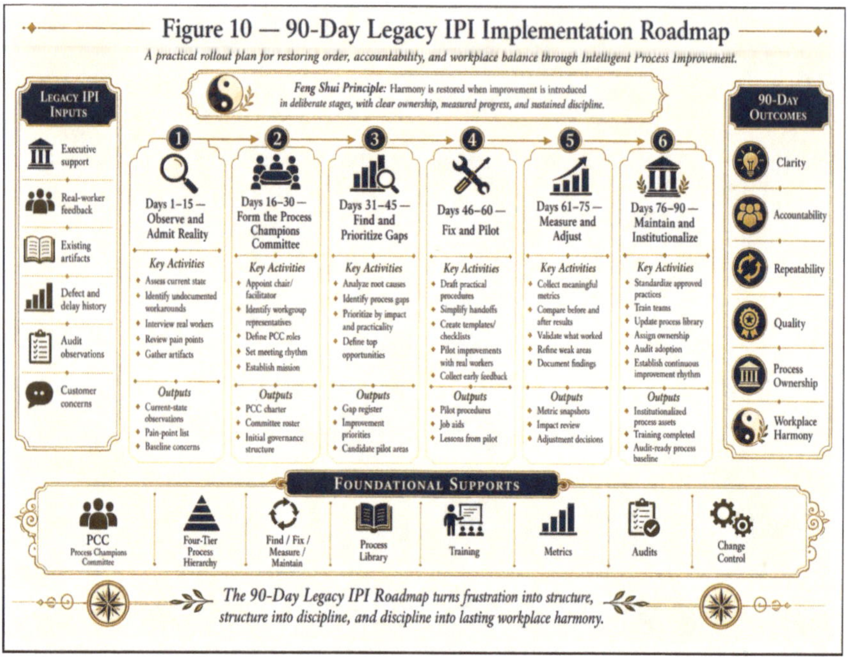

Use this roadmap to turn the 12 steps into an implementation rhythm: observe, form the PCC, find and prioritize gaps, fix and pilot, measure and adjust, then maintain and institutionalize.

Seven Critical Principles of IPI

1. Intelligently

IPI believes intelligent process decisions should be made as close to the work as possible. Executives see the strategic picture, but real workers see the daily flow, the blocked handoff, the missing artifact, the awkward template, and the workaround everyone pretends is not happening.

Legacy Reminder

The people closest to the work usually know where the process breaks first.

2. Incrementally

IPI improves in manageable increments. It avoids the big-bang process initiative that collapses under its own kickoff deck. The organization should improve enough to learn, measure, and stabilize before moving to the next improvement.

Legacy Reminder

A small improvement that survives real project pressure is better than a grand rollout that collapses after kickoff.

3. Measurably

IPI insists that every serious improvement be measured. The measure should include delivery outcomes, quality signals, process compliance, and the human effects of process: frustration, empowerment, clarity, and harmony.

Legacy Reminder

If the number does not help the team decide, correct, or improve, it is decoration.

4. Harmoniously

Harmony is not softness. Harmony is disciplined workplace alignment. Roles are clear, expectations are visible, handoffs are understood, and people can do meaningful work without constant defensive posture.

Legacy Reminder

Harmony is not the absence of discipline; it is discipline arranged so people can work without unnecessary friction.

5. With Empowerment

IPI empowerment means giving real workers the authority, evidence, and process structure to improve their own environment. Accountability becomes positive when people help define how the work will be judged.

Legacy Reminder

Accountability empowers workers when they help define how success is measured.

6. By Real Workers

IPI never forgets the analysts, testers, developers, project managers, configuration managers, trainers, process librarians, auditors, support teams, and customer-facing staff who live inside the process every day.

Legacy Reminder

A process designed without real-worker input will eventually be redesigned by real-worker workarounds.

7. Against Frustration

Frustration is not merely a feeling. It is process data. Rework, escalations, late approvals, unclear ownership, unnecessary meetings, and heroic weekends are symptoms of blocked workplace flow.

Legacy Reminder

Frustration is process evidence; repeated frustration usually points to blocked flow, unclear ownership, or missing artifacts.

IPI How-To: Legacy Implementation Questions

How does IPI relate to CMMI?

IPI can work with CMMI because it is an implementation approach, not a competing maturity model. CMMI gives structure; IPI translates structure into practical workplace behavior.

How does IPI change the relationship between executives and real workers?

IPI gives executives visibility into solutions instead of only escalated problems. Real workers find, fix, measure, and maintain improvements, then report evidence upward through the PCC.

Why are Business Analyst skills valuable to the PCC?

Business Analysts know how to elicit, map, analyze, prioritize, and document information. Those skills apply naturally to process improvement because processes are also requirements for how work should move.

How does IPI keep custom processes brief?

IPI favors short, direct process assets: worksheets, checklists, flow diagrams, glossary entries, local procedures, and project practices that can be read and used before the worker forgets why they opened the document.

Why is process mapping required?

Process mapping is the visual roadmap. It exposes gaps, handoffs, delays, unclear ownership, and the places where work leaves one department and enters another without a passport.

How does IPI prevent over-documentation?

IPI uses best-practice sandboxes, process expiration dates, pilot cycles, and process review to keep the library useful. Not every good idea deserves a mandatory form.

How does IPI handle executives bypassing the process?

IPI uses executive waivers. If leadership needs to bypass a policy or procedure, the exception should be visible, justified, time-limited, and stored in the repository.

How does IPI stay fresh?

Processes should have owners, review dates, measurements, training updates, and retirement rules. A stale process is workplace clutter.

How does IPI stay flexible?

Steps can be marked required, optional, or project-defined. Flexibility is allowed, but it should be intentional, visible, and accountable.

How does IPI address the whole workplace?

IPI engages intelligence through real-worker knowledge, surroundings through Feng Shui-inspired flow and balance, and values through accountability, harmony, empowerment, and pride in the work.

Pocket Tools and Worksheets

Tool 1 — IPI 12-Step Readiness Card

Quick Check	Yes / No	Notes / Owner
We have named the real workers affected by this process.	☐ Yes ☐ No ☐ Partial	
We have documented the current as-is reality without blame.	☐ Yes ☐ No ☐ Partial	
We have drawn the process flow and marked the gaps.	☐ Yes ☐ No ☐ Partial	
We have assigned PCC ownership or review.	☐ Yes ☐ No ☐ Partial	
We have defined how the improvement will be measured.	☐ Yes ☐ No ☐ Partial	
We have planned how the process will be maintained or purged.	☐ Yes ☐ No ☐ Partial	

Tool 2 — Process Improvement Suggestion Slip

Suggested improvement	
Current frustration or blockage	
Affected roles	
Existing artifact or missing artifact	
Proposed fix	
Suggested metric	
Submitted by / date	

Tool 3 — Mini Harmony Dashboard

Area	Current Signal	Desired Balance	Action
Clarity	High / Medium / Low	What should harmony look like?	Owner / next action
Accountability	High / Medium / Low	What should harmony look like?	Owner / next action
Frustration	High / Medium / Low	What should harmony look like?	Owner / next action
Empowerment	High / Medium / Low	What should harmony look like?	Owner / next action
Handoff Flow	High / Medium / Low	What should harmony look like?	Owner / next action
Process Usefulness	High / Medium / Low	What should harmony look like?	Owner / next action

Tool 4 — Process Purge Review

Quick Check	Yes / No	Notes / Owner
This process still supports a current business or project need.	☐ Yes ☐ No ☐ Partial	
This process has an active owner.	☐ Yes ☐ No ☐ Partial	
This process is used by the people it affects.	☐ Yes ☐ No ☐ Partial	
The template or artifact is still necessary.	☐ Yes ☐ No ☐ Partial	
The process has not been bypassed so often that the waiver became normal.	☐ Yes ☐ No ☐ Partial	
The process improves clarity, quality, accountability, or harmony.	☐ Yes ☐ No ☐ Partial	

PCC First Meeting Checklist

- Confirm PCC chair/facilitator
- Identify standing members
- Identify rotating real-worker representatives
- Approve meeting rhythm
- Create process suggestion intake method
- Identify first three process pain points
- Assign process library owner
- Define waiver tracking method
- Schedule first process audit/review

Glossary of Legacy IPI Terms

PCC — Process Champions Committee

The Process Champions Committee is the standing IPI governance group responsible for guiding, reviewing, approving, auditing, and sustaining process improvement across the organization. In the legacy IPI model, the PCC gives process improvement a home, a voice, and a decision-making body made up of process leaders and real-worker representatives. Its purpose is to keep process discipline balanced, useful, and accountable.

FFMM — Find, Fix, Measure, Maintain

FFMM is the core IPI improvement cycle. First, the organization finds the process problem, gap, friction point, or source of frustration. Then it fixes the issue through a practical process improvement. Next, it measures whether the fix actually improved the work. Finally, it maintains the improvement through ownership, training, audits, library updates, and continued review.

As-Is Process

The as-is process is the current way work is actually being performed, whether formally documented or informally practiced. In IPI, capturing the as-is process requires honesty. The goal is not to punish workers for shortcuts or undocumented workarounds. The goal is to understand reality before designing improvement.

Process Library

The process library is the organized repository where approved process assets are stored, maintained, versioned, and made available to the organization. It may include procedures, templates, checklists, process maps, forms, standards, training material, lessons learned, and audit evidence. In legacy IPI, the process library is the organization's process memory.

Waiver

A waiver is a documented exception to an approved process, policy, procedure, or required step. A waiver should be justified, approved, time-limited, and stored for future review. In IPI, repeated waivers are treated as warning signs that the process may be too rigid, unclear, outdated, or misaligned with the real work.

Harmony Dashboard

A harmony dashboard is an IPI measurement tool used to track whether the process environment is improving the workplace, not merely meeting deadlines. It may include indicators such as clarity, frustration, empowerment, accountability, handoff flow, quality, rework, and process usefulness. Its purpose is to measure whether process improvement is creating balance and harmony for real workers.

Process Purge

A process purge is the deliberate review and removal of outdated, unused, duplicated, burdensome, or low-value processes and artifacts. In IPI, purging process overload is not anti-process. It is disciplined process housekeeping. The goal is to keep only the processes that continue to serve the work, the worker, and the workplace.

Real Worker

A real worker is any person who performs, supports, manages, tests, documents, maintains, audits, or depends on the actual work of software development. This includes analysts, testers, developers, project managers, configuration managers, trainers, process librarians, auditors, support staff, and customer-facing teams. In IPI, real workers are central because they know where the process breaks before the dashboard admits it.

Process Artifact

A process artifact is any documented output, record, tool, or evidence created or used as part of the process. Examples include requirements documents, process maps, checklists, signoff sheets, test plans, defect logs, metrics reports, change requests, waiver forms, audit findings, and lessons learned. Artifacts help create traceability, accountability, and organizational memory.

Executive Waiver

An executive waiver is a documented exception requested or approved by executive management when leadership chooses to bypass, alter, or accelerate a required process. In IPI, executive waivers are important because accountability must apply upward as well as downward. If workers must follow the process, executives must document when they choose not to. This preserves fairness, visibility, and process credibility.

Final Words to the Real Worker

This quick reference guide returns to the original IPI promise: process improvement should make work better for the people who do the work. It should not bury workers under documents, frighten them with audits, or impress executives with diagrams that never touch the actual project floor.

Use these 12 steps as a practical reminder. Find what is blocked. Fix what can be improved. Measure whether the improvement truly helped. Maintain the process only as long as it serves the work. Purge what has become clutter. Teach what matters. Protect the real worker. Restore harmony one process object at a time.

Coming Next in the Series

This quick reference guide supports Feng Shui for Intelligent Process Improvement: Legacy Principles of Structured Software Development and points forward to Process Intelligence: The Missing Link Between Agile, DevOps, and AI. The legacy guide preserves the roots; the next book carries the principles into the modern operating layer.

www.ingramcontent.com/pod-product-compliance
Lightning Source LLC
LaVergne TN
LVHW051131210926
848482LV00001B/51
* 9 7 9 8 9 9 5 8 7 2 5 6 6 *